THE ESSENTIALS OF PACKAGING

THE ESSENTIALS OF PACKAGING

A GUIDE FOR MICRO, SMALL, AND MEDIUM SIZED BUSINESSES

SOLA SOMADE AND TUNJI ADEGBOYE

THE ESSENTIALS OF PACKAGING
A GUIDE FOR MICRO, SMALL, AND MEDIUM SIZED BUSINESSES

iUniverse books may be ordered through booksellers or by contacting:

iUniverse
1663 Liberty Drive
Bloomington, IN 47403
www.iuniverse.com
1-800-Authors (1-800-288-4677)

Because of the dynamic nature of the Internet, any web addresses or links contained in this book may have changed since publication and may no longer be valid. The views expressed in this work are solely those of the author and do not necessarily reflect the views of the publisher, and the publisher hereby disclaims any responsibility for them.

Any people depicted in stock imagery provided by Getty Images are models, and such images are being used for illustrative purposes only.
Certain stock imagery © Getty Images.

ISBN: 978-1-5320-4378-9 (sc)
ISBN: 978-1-5320-4379-6 (e)

Library of Congress Control Number: 2018902544

Print information available on the last page.

iUniverse rev. date: 04/26/2018

Contents

Preface

Packaging remains a big challenge in the world of commerce and industry, where competition, largely driven by globalisation, has become a buzzword in recent years. Any company or nation that wants to remain relevant or seeks to be part of the global players in commerce and industry cannot afford not to pay adequate attention to its packaging needs. This is because no matter what goods are to be produced, packaging remains the most effective medium through which such goods can be made available to the consumer wherever he resides.

Therefore, The Essentials of *Packaging: A Guide for Micro, Small, and Medium-Sized Businesses,* has been written to guide manufacturers or anyone who has any marketable product to package. As the book's title shows, some emphasis has been focused on MSMEs that are at a disadvantage for various reasons, among which is their lack of financial muscle to engage full-time packaging professionals to handle their packaging needs on a routine basis.

Why emphasis on MSMEs? Whenever the authors were called to make presentations at seminars, workshops, etc., where MSMEs were participants, packaging came out clearly as a huge problem. Many, who had fully developed products, were in a dilemma as to how to package them. Some already have products wrongly packaged. Almost all sessions we had presided over ended up with packaging clinic sessions during which some time was spent to assist in sorting out various packaging issues being faced by this sector.

Considering the role of the MSMEs as the foundation or the pillar on which the industrial growth of any nation depends, it will be suicidal to allow the operators in this vital sector to remain helpless in an area so crucial to their survival and growth. Any nation that neglects the vital needs of her MSME operators is only building its economy on quicksand.

The book is not aimed at making MSME operators or others packaging professionals. It is just a quick guide that points them in the right direction of where to seek help. It is meant to enlighten this sector on the basics of packaging that will allow them to select a packaging that will be successful in the market and sell their product.

The book opens with brief comments on the MSMEs and the vital role they play in the economy of any nation. The next chapter defines packaging and its main functions as they affect consumer packaged goods. The next five chapters examine the major packaging media or materials, their strengths and weaknesses and areas of application. The last three chapters briefly discuss packaging printing and decoration, packaging specifications, and packaging and the environment.

The authors have deliberately kept the book short and simple. The language has been made simple and technical terms have been limited to where they are absolutely necessary for effectiveness. In such cases, a glossary of packaging terms has been created in the Appendix—to explain in simple language the meanings of such words.

We have written this book with the hope that it will give some relief to the MSME operators in the area of packaging. The book should also be a good companion to anyone who has anything to do with the packaging or repackaging of any finished, semi-finished goods, or unprocessed raw goods like fruits and vegetables for the marketplace.

We hope this book is found useful by the MSMEs.

Sola Somade

Tunji Adegboye

May 2nd, 2017

The small and medium scale enterprises: Who are they?

Definition

MSMEs is the recognized abbreviation for the micro, small and medium-scale enterprises. The definition of MSMEs varies among different countries of the world, depending on each country's level of development. Usually, the definition is based on diverse considerations, such as the revenue levels, e.g. annual turnover, level of profit, and number of employees. For example, the Small Business Administration (SBA), the government department in the USA responsible for the MSMEs, uses the term "size standards" to indicate the largest a concern can be to still be considered as a small business and, therefore, able to benefit from small business-targeted funding.

On the other hand, the European Commission adopted a recommendation 2003/361/EC on May 6, 2005, published in OJL 124 of 20/5/2003, p. 36. Here, the Commission has a third category called the Micro Enterprises. A micro enterprise has a headcount of less than 10 and a turnover or balance sheet total of not more than Euro 2 million. A small enterprise has a headcount of less than 50 and a turnover or balance sheet total of not more than Euro 10 million, while a medium enterprise has a headcount of less than 250 and a turnover of not more than Euro 50 million. All member states are expected to apply this definition for consistency in dealing with policies regarding these three groups of enterprises.

In the developing countries, the definition for MSMEs differs. The term applies to companies or enterprises with head count of less than 100, and these enterprises are relatively smaller in monetary terms than their counterparts in the developed countries.

Importance of MSMEs

Globally, MSMEs have been recognized to play a major role in promoting grassroot economic growth, particularly in the developing countries. They are the backbone of any nation's economy. They are indeed the pillars on which the economies of nations rest. The MSMEs constitute the bulk of the industrial base and contribute significantly to the Gross Domestic Product (GDP) of nations. In many economies, they constitute over 90 percent of the total numbers of enterprises. They are vital instruments for the increase in the industrial production, as well as exports. They play an important role in poverty alleviation programmes of governments through employment generation.

As MSMEs become stronger in their contribution to nations' economies in terms of employment creation, higher contribution to the GDP, etc., globally, there are increased and deliberate policies and legislations by the various governments aimed at nurturing and strengthening the MSMEs as engines of economic growth.

Some Challenges Facing the MSMEs in the Developing Countries

1. Many lack capital adequacy to support their operations. This is because many of them start as one-man businesses and they have limited access to conventional sources of funds, like banks and other financial institutions.
2. Poor and inadequate infrastructural facilities, such as power, water, and roads not only hinder their progress and performance, but also add so much to the cost of running their businesses.

3. Government support is at best feeble, weak, and can hardly be relied upon for sustainable operations.

4. Most of the machines and equipment they use are locally fabricated and are rather crude, hence efficiency and productivity suffers.

5. Though labour abounds in many developing countries, they lack the relevant managerial competence and technical know-how to make the much-needed impact on the enterprises.

6. From the point of view of pricing and product quality, many MSMEs in the developing countries cannot compete in the global market, particularly in the advanced countries of the West.

7. Satisfactory product packaging remains a sore issue for many MSMEs, which is the reason for writing this small book.

8. As vital as the MSMEs are to the development of any economy, they are the most vulnerable to the vagaries of economic and political instability of any nation, as we are presently witnessing in many parts of the world, and the reasons for this are quite obvious.

It is important to know that production creates the wealth of a nation, while other functions assist with its distribution. For a nation to grow industrially, government must truthfully and faithfully be committed to assisting the MSMEs to achieve their goals. The truth as Noah Webster put it is, 'You may reason, speculate, complain, form mobs, spend your life railing at Congress and your rulers, but unless you import less than you export, unless you spend less than you earn, you will be eternally poor.' The solution to poverty is high productivity and the ability to compete globally. Vibrant MSMEs hold the ace.

Packaging and its functions

What is Packaging?

Packaging can best be described as a coordinated system of preparing goods for transport, storage, distribution, retailing, and delivery to the ultimate consumer in a sound condition at an economic cost. Packaging is a service function. It cannot exist on its own. There must be a product; otherwise, there is no need for packaging. Packaging is a dynamic, scientific, and artistic business that cuts across many disciplines, such as science, engineering, marketing, arts, and social sciences.

Packaging Functions

From the definition given above, some insight into the functions of packaging is obvious. The major functions are:

1) **The Containment Function**: The product must be contained within the packaging material. If the containment function fails, the protection and preservation function has no chance of succeeding. In order to achieve the containment function, the physical attributes of the product must be known so that the appropriate choice of packaging material can be made.

2) **Protection and Preservation**: This is the primary function of packaging. Every product is susceptible to one form of damage

or spoilage or the other. Therefore, it is the duty of packaging to prevent or mitigate damage or deterioration. Sources of spoilage could be from the nature of the product, the environment, handling during storage and distribution. No matter what, packaging must be able to take care of all likely threats to the wholesomeness of the product throughout the supply chain. Packaging demand in terms of protection and preservation varies from product to product depending on how fragile, how hygroscopic, how fatty, etc. the product is. Damaged containers are unacceptable; likewise, the damaged goods inside them. For example, crushed bakery products like cakes, biscuits, and cookies are unsightly and unacceptable to the consumer. Preservation often refers to the extension of the life of food beyond its natural life, or maintenance of sterility in food or medical products. Packaging must prevent product failure throughout the shelf life of the product.

3) **Information Function**: Packaging must communicate the appropriate information about the product to the consumer. With modern retailing and shopping activities, the information carried by the package must include brand name of the product, the manufacturer's name and address, with the telephone number, e-mail address, and the website, ingredients' list, direction for use, weight/volume declaration, manufacturing date, batch number, Best Before date, and handling and cautionary details. Finally, a bar code that carries electronic data that assists supermarket chains to keep track of the stock level and movement of goods among others has become compulsory on consumer goods. The information on packaging has been largely responsible for the success of self-service in today's modern supermarkets and hypermarkets.

4) **The Selling Function**: No doubt, the information provided on the package assists in selling the product. However, packaging presentation helps in no small measure in the selling activity. Impressive eye-catching designs in graphics, texts, or logos on containers attract the customer's attention to the products, and this leads to a lot of impulse buying, at least to the first time buyers. Therefore, a good graphic design with effective use of

colour combination is a strong selling point. Also, the structural design of the packaging material, in various shapes, enhances the pack aesthetics, and this influences the consumer perception of the product. But we must remember that no matter the quality or the sophistication of the package and its graphics, if the quality of the content is poor, you can deceive the consumer only once. What sustains a brand and makes it grow is not the first purchase, but the subsequent repurchases by the satisfied and loyal customer who will always come back.

5) **Storage and Transportation**: Storage and transportation of many processed goods would be impossible without packaging. Just imagine how difficult it will be for products like milk (powder or liquid), beverages (alcoholic or non-alcoholic), detergents, and cereals, to be stored and transported safely without packaging. Packaging does not only protect and preserve, it also makes handling of products throughout the supply chain possible and convenient. Packaging makes the arrangement of goods on pallets and stacking of the same in warehouses using various handling equipment possible. Also, transporting the goods within and across borders by the various transportation modes would have been impossible but for packaging. In short, without packaging, international trade would have been very weak, low, and unsustainable.

6) **Advertising Medium**: Packaging has emerged as the most effective medium through which many packaged goods are advertised. It is now very common to see products being presented to the consumers in the comfort of their homes on the TV in their beautifully designed packages. Also, many promotional and product re-launch activities are executed through modified and redesigned packages or in new packages.

7) **Convenience:** One of the purposes of packaging is to make life comfortable for the consumer. However, some products depend entirely on the promise of convenience offered by their packaging materials for their success. The convenience factor or promise could be premised on efficiency in dispensing, economy in use, easy disposal, and/or storage. Among the products that fall into this

category are aerosol insecticide containers, perfumes with special dispensers, shampoos with special pouring or dispensing spouts, precooked foods, and do-it-yourself kits. The list is endless. As the consumer's demand for sophistication grows and technological innovation improves, more and more convenience features are constantly being built into packaging materials/containers.

8) **Tamper Evidence**: A closure is a vital and crucial part of any packaging container. A good closure must reveal that unlawful access to the content of a container has occurred. Therefore, adequate telltale features must be built into the closure system so that any unauthorized access is revealed. Apart from the tamper-evident type of closure, there is the child-resistant type, which has been designed to make it difficult for young children to open the container while it is still easy for adults to open.

Levels of packaging

Packaging materials can be categorized into three levels or groups. These are the primary, secondary, and tertiary levels.

a) **Primary Packaging**: These are the packaging materials that are in direct contact with the products and are the first line of defence against the environment. These materials must therefore be compatible with the products and must not in any way react or interact with them. Examples of primary packaging are sweet wrappers, beverage cans, detergent cartons, laminated/flexible bags, cosmetic jars, beer bottles, and several others.

b) **Secondary Packaging**: These are the collating units that hold or contain some units of the primary containers. In this role, secondary packaging is recognized as the distribution packaging. The most popular secondary packaging is the ubiquitous corrugated case. Others are plastic crates, shrink-wrapped trays, etc. Essentially, secondary packaging unitizes and compliments the primary packaging in various ways. Apart from the fact that they make palletisation, storage, stacking, and transportation of goods possible, they protect the primary containers from physical damage, which could also affect the product inside them. In some situations, the secondary packaging does not play the role of a shipping container or distribution packaging. In such cases, they are just seen as the next layer of packaging to the primary packaging. Examples of such situations are a display box containing a tube of cream or toothpaste, a display box holding a glass jar of body cream, etc.

c) **Tertiary Packaging**: This is usually, but not always, the distribution packaging. But in a situation where the tertiary packaging is not the shipping/distribution packaging, then it means it does not exist as part of the total package. It means there are only two packaging layers—the primary and the secondary, with the latter constituting the distribution packaging. The most common tertiary packaging is the corrugated fibreboard shipping container. It can be difficult at times to make a distinction between primary, secondary, and tertiary packaging. For example, we have seen situations where the corrugated box has featured as primary packaging, secondary packaging, and tertiary packaging.

Paper and paperboard packaging

Paper and paperboard are among major modern packaging media. The raw material for paper and paperboard manufacturing is the same - cellulose fibre, which is derived from wood. Other sources of fibre apart from wood are cotton, bamboo, straw, and "flax and esparto."

The generation or extraction of fibres from wood and other sources of fibre is beyond the scope of this book. Also beyond the scope of this book is the production of paper and paperboard from the fibres extracted from the wood and other sources of fibre. The general aim of this chapter is to bring into focus the areas where paper and paperboard feature as packaging materials within the MSMEs' activities and in product manufacturing and packaging.

Let us first clarify the difference or distinction between paper and paperboard. Essentially, the difference is in the basis weight—that is, the gramme weight per unit area of the material. The International Standards Organisation defines paper-based materials weighing 250 grams per square metre (gsm) or more as paperboard. Anything below that is regarded as paper. The implication of this basis weight demarcation is that paper (< 250 grams per square meter) (gsm) shall be regarded as flexible packaging, while paperboard ≥ 250 gsm falls into the category of rigid packaging.

Paper and paperboard packaging can be broadly categorized into three groups and these are paper, folding cartons, and corrugated boxes. Each of these will now be discussed with a view to highlighting their relevant areas of use in packaging.

1. **Paper**: Paper is produced to the lowest basis weight among the three groups and therefore falls into the flexible packaging group. And within the paper industry itself, there are two categories of uses—the packaging and the non- packaging uses. The non-packaging uses include printing and writing papers for books, magazines, stationery, and newspapers, where the main input is the newsprint.

There are many types of packaging papers. Until recent times, paper was the main material used for labels. Most beer bottles still carry well-printed paper labels. People hardly regard envelopes in their various colours, sizes, and designs as packaging. But they are. Why? They contain and protect the contents until they and the contents are delivered to the recipients. Another area where paper is used as packaging is the wrapping of various consumer goods such as soaps, foods, confectionery, snacks, etc.

In most cases, the paper must have been specially treated to enhance its functional performance one way or the other. This is because untreated or unmodified paper is very porous and is hardly suitable for most packaging applications without such treatments or modifications.

There are also specialty papers, which are used in food and confectionery wrapping. One of them is the glassine paper, which is smooth and glossy. Another is the vegetable parchment paper, which is specially treated to make it grease and oil resistant. It is used to wrap fat-based foods such as butter and margarine.

Another area where paper is used as packaging is in bulk shipment, where paper is converted into bags or sacks, with or without inner liners, for bulk shipment of goods. The sack could be made of a single high caliper gauge paper or two or more low gauge plies of mostly unbleached kraft paper. Some of the goods that are mostly packed in these sacks or bags are cement, fertilizer, animal feed, agricultural produce, and chemicals. Plastic bags, woven and unwoven, largely because of their ability to withstand adverse weather conditions, are now gradually displacing the paper sacks.

2. Folding cartons: Boards used for the manufacture of folding cartons are made on board-making machines from pulps generated from wood fibres. There are different types of carton boards. Among the most common boards for folding box manufacturing are:

i. **Solid Bleached Board**: This is made entirely of bleached virgin chemical pulp completely free of recycled material. It is a high quality board noted for its performance and high presentation, particularly when printed using lithographic or gravure printing processes. Folding cartons produced using this kind of board are suitable for use in the pharmaceutical and cosmetic industries.

ii. **Folding Box Board or Duplex Board**: This board consists of three layers, *viz* a top layer/liner of bleached chemical pulp, a middle layer of mechanical pulp, and a back layer of bleached or unbleached chemical pulp. The top layer is highly polished (machine glazed) and smooth in preparation for excellent print quality. The cartons produced with this board are widely used in food packaging.

iii. **White Lined Chipboard**: This board is made of a top layer of bleached chemical pulp, middle layer of recycled waste paper, while the back layer could be made of virgin pulp or recycled waste, depending on what the folding cartons will be used to package. Most of these cartons are either used directly as primary packaging for such diverse products as detergent powders, non-edible starch, soap tablet, or as display boxes (secondary packaging) for such products as cereals, cocoa-based beverages, toothpastes, tea bags, etc., where there are already protective inner packaging materials that contain the products. This "bag-in-box" concept is common where a product packed in a flexible bag is seen to be susceptible to mechanical damage.

There are other types of board in use but the above named three dominate the paperboard-packaging scene. However, one must not overlook the moulded paper pulp in packaging, which was, and is still, being used in egg packaging and distribution. Here again, moulded plastic options are already giving moulded paper pulp a good challenge.

Folding Carton Production

Once the type and grade of board has been decided—the choice that in most cases depends on product and the weight/volume of the product to be contained or packed in the carton—the next stage is the dimensional specification. For a given volume of carton, the dimensions usually in the order of length, width, and height can vary a lot. The dimensions give shape to the carton. Deciding or choosing the dimensions and flap style is essentially what the structural design is all about, and many carton manufacturers now use Computer Aided Design (CAD) system to achieve this goal. Having decided the design, a plain sample blank is made for the customer's approval. This is followed by the graphic design stage, which can be a long process, particularly if one is dealing with a complex multicolour printing design. The choice of the printing method is dictated by the expected quality of printing, volume or quantity to be printed, and available printing facilities on ground.

The manufacturing of folding cartons starts with the printing of the boards. The board can be printed as a continuous web or as sheets by any of the printing processes discussed in the chapter on package printing.

The next step after printing and drying is the cutting and creasing stage. The purpose of cutting and creasing is to enable the carton blank to be formed from the flat sheet or reel of the carton board. This is done on a die-cutting forme, fitted with steel cutting knives and blunt steel creasing rules. The "cuts" determine the boundaries of each carton blank, while the creases define the vertical and horizontal fold lines. The cutting and creasing operation also makes the separation of the blanks and the waste sheet possible through the process of stripping. After stripping, gluing of side seams follows and this can be done manually or on gluing machines, depending on the volume and complexity of the job to be done. Finally, packing of the folding cartons into shipping containers in bundles of twenty-five or fifty cartons for delivery to the users completes the manufacturing process. The quality of the final carton depends on the attention paid to each of the above listed stages of carton making.

3. Corrugated shipping boxes

Corrugated fibreboard box or case is the most common shipping/distribution container in use today, accounting for not less than eighty percent of all shipping containers worldwide. They come in many structural styles, the 'Regular Slotted Container' (RSC) being the most common.

The board from which a corrugated box is constructed is made of kraft liners and fluting/corrugated medium. Generally, there are four types of corrugated board, depending on the number of liners and fluting medium in the construction. The four types are:

i.   Single face corrugated board, which consists of one linerboard and one flute.
ii. Single wall corrugated board, which consists of two liners and one flute. This is the most common board in use and it consists of an outer liner, fluting medium, and an inner liner.
iii. Double-wall corrugated board, which consists of three liners and two flutes.
iv. Triple-wall corrugated board, which consists of four liners and three flutes.

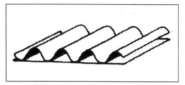

Single face corrugated board

Single-wall corrugated board

Double-Wall Corrugated Board

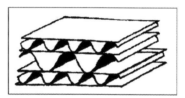

Triple-Wall Corrugated Board

Source: ITC Packit - Packaging Design (2005)

Corrugated boards
Figure 3.1

The liner layers can be made from pure kraft, jute, or test liner, and they are made in different thicknesses and basis weights, the latter specified in grams per meter squared (gsm) or in pounds per thousand square feet (MSF). The flutes are made of semi-chemical board material. Like the liners, they are quoted in weights per unit area. But only about three nominal values are recognized worldwide, and these are 112 gsm, 127 gsm, and 150 gsm in metric units and 23 MSF, 26 MSF, and 31 MSF, respectively in imperial units. But unlike liners, flutes come in four main different configurations. These are designated as A, B, C, and E flutes. The basic differences are in the number of flutes or corrugations per unit length (foot or meter) and the height of the flutes. Nevertheless, these basic differences have profound influence on the performance characteristics of the corrugated boxes.

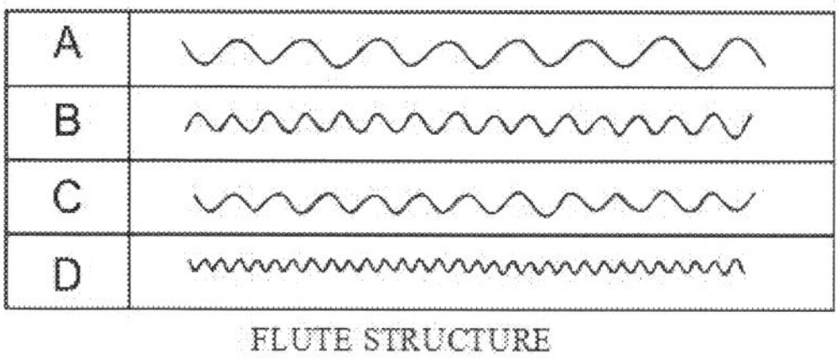

FLUTE STRUCTURE

Source: Handbook on Procurement of Packaging (1989)

Flute Structure
Figure 3.2

The final point to make on corrugated boxes is the specification. Normally, the grade of the board to be used for the box must be specified. This is done by stating the makeup of the corrugated box in terms of the material substances of the outer liner, the flute, and the inner liner in that order. For example, a 300K/112B/200K corrugated board means the outer liner is a pure kraft of 300g/m^2 grammage, the flute is a semi-chemical material of 112g/m^2 grammage, and the inner liner is a pure kraft of 200g/m^2 grammage. The other parameters that must be specified are the case/box dimensions with relevant tolerances. Specified cases dimensions are always internal as measured from one scoring line to the adjacent one and are in the order Length (L), Width (W), and Depth (D)/Height (H). Case printing based on the approved artwork must be specified, as well as the mode of securing the case's manufacturer's joint or lap. This could be done by stitching or by glue application.

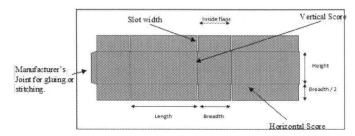

**BLANK FOR REGULAR
SLOTTED CASE (RSC)**

Source: Handbook on Procurement of Packaging (1989)

Figure 3.3

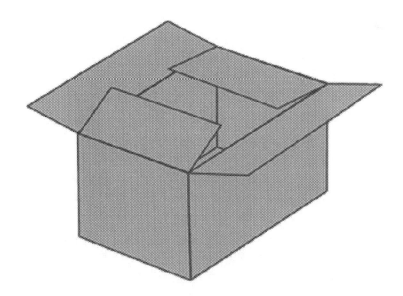

Source: Handbook on Procurement of Packaging (1989)

An erected corrugated case
Figure 3.4

The functions of the corrugated boxes are:

i. It must contain a specified number of retail units of packed products.

ii. It must be easy to palletize efficiently on a standard pallet for storage and stacking.

iii. It must be strong enough to withstand a reasonable period of storage and stacking.

iv. It must provide mechanical protection to the retail units it contains during storage and distribution.

v. It must carry the appropriate information for all interested stakeholders.

vi. It must be cost effective and cost no more than is absolutely necessary to do its job.

Metal packaging

Introduction

Metal is a major packaging material for food and beverages. Other products such as paints, petroleum products, etc., are packed in various metal containers. However, in recent times, most paint and petroleum products are now packed in moulded plastic containers.

Metals usually employed in packaging are iron in the form of steel, tin, chromium, and aluminium. These metals are used in a combined or alloyed state with one another to enhance strength, ductility, corrosion resistance, and other functional attributes. The various combinations of the above metals result in the following well-known metal packaging materials.

1. Black Plate: Uncoated Steel
2. Tin-free Steel: Electro-coated Chrome-coated Steel
3. Tinplate: Tinplated Steel
4. Aluminium alloy, Slugs or Sheets
5. Aluminium Foil

While the first two are hardly in use for most fast-moving consumer goods (FMCG), the last three constitute the core metal packaging materials. Therefore, two out of the three (tinplate and aluminium alloy), which are used mostly as rigid metal containers, will be discussed in this chapter, while aluminium foil will be addressed in the chapter on flexible and laminate packaging.

Tinplate

This is a thin low carbon steel (0.15-0.30 millimetre mild steel) coated with a layer of tin on both sides, either by a *dipping* process or lately by an electrolytic process. While the dipping process allows the same level of tin coating on both sides of the steel sheet, the electrolytic process has the flexibility of allowing differential tin coating levels. Tinplate sheets can be made into three-piece (3-Piece) or two-piece (2-piece) cans using different making technologies. Before discussing the can-making processes available, let us look at the advantages and disadvantages of tinplate as a packaging material.

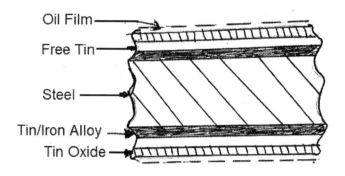

Oil Film
Free Tin
Steel
Tin/Iron Alloy
Tin Oxide

Tinplate Cross-section

Figure 4.1

Advantages of tinplate

1. It offers superior (100 percent) protection/barrier against moisture, light, gases, etc.
2. It is strong, stiff, and ductile.
3. It functions at extreme temperatures, hence products in metal cans can be subjected to high temperature pasteurisation/sterilisation as required in some food products, e.g. tomato puree.
4. The surface is suitable for external decoration (printing) and for internal protective coating with organic materials/compounds (internal lacquering).
5. It is easily convertible into cans on very high-speed machines.
6. It is opaque and therefore suitable for products, which are sensitive to light.
7. Tinplate is easy to recycle.

Disadvantages

1. Metal cans are highly susceptible to corrosion, particularly in a humid and high temperature environment.
2. It is very expensive compared to other packaging materials.

3. It has sharp edges and therefore needs to be handled with care to avoid personal injury.

Tinplate Can Making (3-piece Cans)

To manufacture this style of can, a rectangular blank of tinplate is mechanically wrapped around a mandrel to form a tube or cylinder. The body blank is sized to be capable of being formed into one can body. Before making the tube, the body blank is stamped out at the four corners, where specially shaped notches are created so that instead of the interlocked seam being formed over the whole length of the can body, the extreme ends are simply overlapped. During body forming where the two ends of the blank meet, an interlocked seam is normally formed, and this is welded to make it leak-proof. In the past, side seam was achieved by soldering, but this has now largely been abandoned for health and safety reasons. The blank sheets constituting the body cylinder are normally printed and dried before the body forming operation, and printing is usually done using the offset lithographic process.

The formed bodies are transferred to the flanging machine, where the rims (the two ends of the body) are flanged outwards to enable the ends to be seamed on.

The can ends or lids are produced or stamped from similar sheets of tinplate. To prevent leakage, a resilient lining compound is applied to the whole peripheral area of the ends and cured before seaming takes place. The end is then mechanically joined to the body cylinder by a double seaming operation in which two flanges or hooks of the body cylinder and the end are mechanically interlocked. The can maker applies one end, while the other is applied by the food processor after product filling.

Three-piece tinplate cans
Figure 4.2

The 2-Piece Can Making Process

Two-piece containers may be produced using either the tinplate sheet or the aluminium sheet. Two major methods of producing this type of can now exist. The success of the two methods depends on the property of the metals to flow without rupturing. Both the "Drawn and Re-Drawn" (DRD) and "Drawn and Wall Ironed" (DWI) methods produce lighter weight containers without side weld or seam. This contrasts with the three-piece cans.

1. Drawn and Re-Drawn (DRD) Method

Here, a circle is stamped out of usually pre-lacquered metal sheet. The circle is drawn into a cup shape and then redrawn two or more times as necessary before being trimmed and shaped as required. At each redraw, the diameter of the cup is reduced and the height increased. Printing or decoration is done on the flat sheet in a distorted fashion so that when the can is formed, the correct image emerges. Cans produced by this method are used as fish, pet food, and polish containers, among others.

2. **Drawn and Wall Ironed (DWI) Method**

Here also, a flat circular blank is punched out of a coil of metal and at the same time drawn into a shallow cup. The cup is then forced through several rings or dies. It is through this dies that the ironing process takes place. The ironing process consequently reduces the wall thickness, just as the body height correspondingly increases. Just like the 3-piece can, the ends for the 2-piece cans are produced from similar sheets of tinplate. But unlike the 3-piece can, only one end is produced for each 2-piece can and the end is applied by the food processor after product filling. The areas of application for the 2-piece DWI cans are beer and carbonated drinks packaging.

Two-piece Aluminium cans
Figure 4.3

Collapsible Tubes

Collapsible metal tubes can be made of tin, lead, or aluminium, but aluminium accounts for almost ninety nine percent of collapsible metal tubes. In recent times, however, collapsible tubes are also made from plastics and from laminates, which have aluminium foil and plastics film as components. The laminate tube is rapidly replacing the aluminium tube because of its equally good barrier properties. Only the production of collapsible tube will now be described.

Aluminium of high purity is cast and rolled to a predetermined thickness and blanked to provide cylindrical pieces of metal of a defined diameter. These pieces, called slugs, are heat-treated to bring them to the desired metallurgical state for fabrication. The slugs are then lubricated, and a process of impact extrusion converts the slugs in a specially designed press to the tube shape, with a shoulder and nozzle. The tubes are then transferred from the press to an automatic machine lathe, where they are trimmed to the correct length and the nozzle incorporated.

The impact extrusion forces leave the tubes in a work-hardened state. The tubes have to be annealed to remove the stiffness and make them collapsible. The next stage is the decoration or printing, which is usually done by dry offset printing. Depending on the product, the inside of the tubes may need to be lacquered or coated for protection, particularly if the product to be filled is very acidic or very alkaline. Usually for closure, injection moulded continuous threaded plastic caps are used for most aluminium tubes.

Aluminium tubes are used as packaging materials for toothpaste, tomato paste, pharmaceutical preparations, cosmetics, fish and meat pate, and dessert toppings.

Advantages of Metal Tubes

1) They are airtight and offer best barrier to gases, flavours, and water vapour.

2) They exhibit dead fold characteristics.
3) They can be beautifully decorated.
4) They have a wide range of lining options because they can withstand high curing temperatures.
5) Product dispensing is very economical because of the very narrow nozzle.
6) They preserve products for a long period of time due to small product exposure during dispensing and have a no suck-back feature.
7) They are opaque, hence are very suitable for light-sensitive products. Aluminium tubes are used as packaging materials for toothpaste, pharmaceutical preparations, tomato paste, cosmetics, fish and meat pate, and dessert toppings.

Disadvantages of Metal Tubes

They are generally very expensive and may be more so if internally lacquered.

Glass Packaging

Glass is an amorphous inorganic substance fused at high temperature and cooled rapidly so that it solidifies to a vitreous condition without crystallisation. Glass is formed from naturally common elements in the earth's crust. It is therefore one of the earliest packaging materials in use.

A typical formula or composition of clear glass is:
Sand (Silica) – SiO_2 73 percent
Soda Ash - Na_2O 14 percent
Lime –CaO 11 percent
Alumina – Al_2O_3 1 percent

Other minor ingredients that make up the remaining one percent (1 percent) are added to achieve one functional goal or the other. For example, small amounts of selenium and cobalt oxides are added as decolourising agents to achieve maximum clarity, that is, white flint glass. Also, addition of small amounts of iron oxide, carbon, and sulphur will result in amber (brown) glass. This brown glass filters off the ultraviolet (UV) rays from the sun from reaching the inside of the glass container. Such bottles are used for products that are sensitive to UV light. Addition of different amounts of iron oxide, manganese, and chromium oxides results in varying shades of green glass. Other glass colours such as blue and opaque white (Opal) are possible by addition of relevant minor additives.

One important material normally used in glass making is cullet or broken glass recovered from the plant operations or from the trade. Cullet can constitute a substantial proportion of the raw material feedstock in the glass furnace. Its presence in the feed enhances the melting rate and significantly reduces energy consumption in glass making.

Glass Manufacturing

It is not our intention in a book of this nature to go into the details of glass making. It suffices to state that glass making is a very capital-intensive process, which requires high-energy input. It is a continuous process that runs twenty-four hours per day all year round and thrives strictly on volume. The origination cost, which consists of bottle or container design and mould procurement, is not by any means cheap. This explains largely why glass packaging is not easily accessible to the MSMEs like other packaging materials. The molten glass that emerges out of the high temperature furnace, in which the ingredients are melted, is formed into two broad categories of containers by adopting two methods of moulding. The two methods and the types of containers produced by each method are:

I. **Blow and Blow Process**: This is normally used for producing narrow-necked bottles.

    a. In blow and blow forming operations, the gob is fed by gravity through a guide/chute into a blank or parison mould, which is in an inverted position.
    b. The guide is replaced by a parison bottomer and compressed air is blown into the mould to force the glass into the finish section where the bottle finish is formed.
    c. The parison bottomer is replaced by a solid bottom plate, and air is forced through the bottle finish to expand the glass upwards into a bottle parison or pre-form.
    d. The parison is transferred from the blank mould, using the neck ring to hold it, and rotated through 180° and placed into the blow mould, where it is blown to its final shape and size.

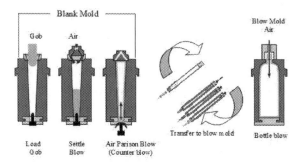

Blow and Blow Process
Figure 5.1

II. **Press and Blow Process**: This is used for the production of wide-mouthed jars. Here, after the gob has been fed or loaded into the blank mould, as in the blow and blow process, the parison is then pressed into shape with a metal plunger entering through the neck ring, rather than blown into shape. The rest of the forming process is exactly the same as stated in (d) above under blow and blow process.

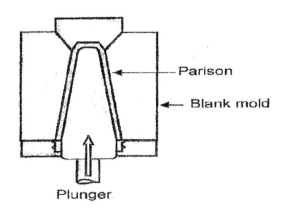

Press and Blow Process
Figure 5.2

Advantages of Glass Containers

1. **Chemical Inertness**: Glass is inert; that is, it will not react with most chemicals, the only exception being hydrofluoric acid.

2. **Barrier Properties**: Its impermeability to moisture vapour and gases is important for long-term storage of products that are sensitive to volatile substances. Glass is a good example of a hermetically sealed pack, provided the closure/container interface is adequately designed to give a perfect seal.

3. **Clarity/Transparency**: This attribute allows product visibility, particularly when this enhances the product image.

4. **Resistance to Internal Pressure:** This property makes glass containers suitable candidates for the packaging of carbonated beverages like soft drinks, beers, and aerosols where internal pressure remains high throughout the shelf life of the product.

5. **Heat Resistance**: Glass is very stable at high temperature, thus making it very suitable for hot filling and high temperature sterilization of products contained in it. This also means glass can be rewashed with steam for reuse.

6. **High Rigidity**: This means container shape/volume does not change or distort under vacuum and under pressure. It also makes it easy to handle during normal production activities and subsequent handling. The resistance to compression also means the outer case (or shipping container) holding a number of bottles can be less rigid as the glass containers load bearing capability is high.

7. **Low Cost:** Glass bottles designed for multiple trips result in low packaging cost since they can be used over and over again before they are disposed of.
   Besides, the raw materials are readily available.

8. **Recyclability/Reuse:** Glass containers can be reused over and over again if properly washed or re-washed. Its high temperature resistance makes sterilization during rewashing possible. Also as cullet, glass can be recycled one hundred percent (100 percent), time without number, without suffering any quality deterioration.

Disadvantages of Glass containers

1. **Fragility/Breakability**: Due to the fragile nature of glass, breakage occurs if it drops on a hard surface.
2. **Weight:** Glass is very heavy, with a specific gravity of 2.5g/cc. In some applications, the weight of the empty glass could be as much as that of the content. With improved technology, light-weighting measures have substantially reduced the weight of glass containers without any reduction in strength. Nevertheless, glass containers still remain the heaviest packaging material, with the attendant consequences on transportation costs.

Container Finish and Closure: The container finish is that part of the glass container specially designed to receive the cap or closure. Since a package is only as good as its closure, in addition to the container quality, the quality of the closure cannot be compromised; otherwise, the proper functioning of the product-package-closure combination cannot be guaranteed. Metal and plastic closures feature most prominently in glass containers.

Glass Container Decoration:

There are three general ways of decorating glass containers.

1. **Use of Labels**: The label material could be paper, aluminium foil, or plastic films. The label can take many forms—e.g. ordinary paper, which will require adhesive application, shrink sleeve that needs heat to activate it. Whichever is the material of choice, all design and application considerations must be carefully undertaken to ensure that the container/cap/label combination are in harmony.
2. **Screen Printing:** This can be used to apply decoration directly to the container surface. Inks must be heat cured to produce durable design.
3. **Etching or Sand blasting:** This design can be done on the glass surface by etching with hydrofluoric acid or by sandblasting.

This is a costly process compared to other methods and is mostly employed in the cosmetic industry, where the cost can be justified.

Glass Applications in Packaging

Glass containers are produced mostly either as bottles or as jars. Other forms of glass containers are vials, ampoules, and syringes, which are normally used within the pharmaceutical industry. The attention here will be more on the conventional bottles and jars. There is no clear distinction between a jar and a bottle. Jars are usually wide-mouthed while bottles are narrow-mouthed. However, it is generally perceived that jars have height/diameter ratio of 2.5: 1 or less, while bottles have a ratio of 3: 1 and above.

Bottles: Products mostly found or packaged in glass bottles are:

i. **Wines and Spirits**: These are high-priced products that naturally have elite value or image and are therefore packaged in materials that convey premium image, like glass, for good complement.

ii. **Beer and Soft Drinks**: Here, the bottles are designed to be able to withstand the high internal pressure resulting from carbonation. Maintenance of high carbonation over the product shelf life cannot be compromised and glass bottles with suitable crown cork closure meet this requirement.

iii. **Fruit Juices and Oils:** Fruit juices and premium oils like olive oil are packaged in glass bottles. In recent times, glass bottles have given way to the use of laminate cartons, Tetra Pack, Pet bottles and other non-glass options.

iv. **Chemicals**: The superior resistance of glass to most chemicals has made it the packaging of choice in the chemical industry. The ubiquitous glassware (flasks, beakers, cylinders, pipette, etc.) in the science laboratories in schools and colleges attest to the popularity of glass as a reliable packaging material for the various organic and inorganic chemicals.

v. **Pharmaceuticals**: Various pharmaceutical syrups like cough syrup, paracetamol syrup, and some other liquid antiseptic preparations are packaged in amber glass bottles.

Finally, of note is the adoption of used glass bottles (reuse) by the small industry operators for the packaging of honey, roasted groundnuts, palm wine, palm oil, etc. in the informal sector of the economies of many developing countries. These operators cannot afford to commission their own moulds for the production of glass bottles for their use due to the very low scale of their operations.

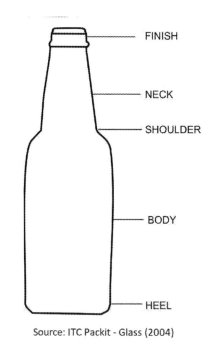

Source: ITC Packit - Glass (2004)

A typical glass bottle
Figure 5.3

Jars: Like bottles, jars are equally very protective of whatever products are packed in them.

i. **Foods:** Glass jars are strong contenders with tinplate cans in the packaging of fruits and vegetables.

ii. **Pharmaceuticals**: Tablets and capsules are packed in jars. To avoid abrasion or breakage of tablets during handling, stuffing materials are often used to fill any empty space at the top of the containers before closure.

iii. **Jams and Preserves**: A high percentage of fruit jams and preserves are packaged in jars using the lug type closure to secure the product.

iv. **Cosmetics**: Hand, body and hair creams, lotions, and pastes are traditionally packed in wide-mouthed jars for easy accessibility and glass jars are competing with plastics here.

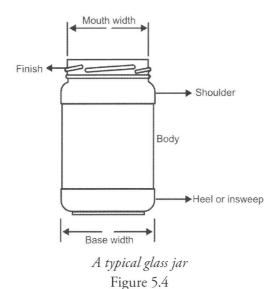

A typical glass jar
Figure 5.4

Rigid Plastics Packaging

Plastics are a broad spectrum of synthetic materials that have the common property of being formed into different geometric shapes and sizes by the application of heat and pressure. Plastics can be broadly classified into two broad groups - thermoplastics and thermosets.

Thermoplastics are fully polymerised plastics, which, once formed, can be melted and formed again and again. Hence, scraps generated during the conversion of thermoplastics can be reprocessed into useful wares.

Thermoplastic materials are numerous and the few that are commonly used in the consumer goods packaging are the polyethylenes (low density-LDPE, medium density-MDPE and high density-HDPE variants), polypropylene (PP) and polyethylene terephthalate (PET) or polyester.

Thermosets on the other hand are plastics that have not been fully polymerised and once heated, formed and cooled, cannot be remelted and reshaped by the application of heat and pressure. Examples of thermosets are phenol and urea formaldehydes. Thermosets are used in the manufacture of some specialty caps/closures, the types normally found in the beauty and cosmetic industry. Their use in the consumer goods packaging is very negligible.

Nearly all of the thermoplastic materials listed above can either be extruded into films (flexible) or moulded into containers (or closures) of various shapes by using the appropriate grade of resins or polymers. The rigid materials so produced can be used as primary or secondary packaging for the consumer as well as industrial goods. The rest of this chapter will address rigid plastics packaging while the flexible packaging will be addressed in chapter 7.

RIGID PLASTICS PACKAGING MANUFACTURING PROCESSES

There are now many methods of producing bottles, containers, trays, crates, and caps from the various plastic resins or polymers. Only the few most popular methods will be discussed here; the many variants and the recent innovative works in these areas are beyond the scope of this book. The major processing methods are:

A. **Injection Moulding**
B. **Blow Moulding**: This has three major variants, namely:
    I.   Injection Blow Moulding
    II. Extrusion Blow Moulding
    III. Injection Stretch Blow Moulding
C. **Thermoforming**

INJECTION MOULDING (IM)

Injection moulding method can be used to produce a wide range of products. Most PET pre-forms needed to produce PET bottles on Stretch Blow Moulding machines are generated by injection moulding process. Most plastic caps, threaded or unthreaded, cups, trays, buckets/pails, crates and pallets are produced using the injection moulding process. The main polymers used are polyethylene, polystyrene, polypropylene and PET (for preforms).

Essentially, the procedure for injection moulding operation is as follows: The polymer or resin is fed through a hopper into the extruder barrel. The polymer is melted here by a combination of heat and friction. A measured amount of the melted but viscous resin is forced through a nozzle into a fully closed matched mould of usually many cavities designed precisely to the shape and size of the final product. The mould has internal cooling system so that the melted polymer may be solidified into the desired shape. When the moulded package (cap, bucket, tub, etc.) is sufficiently cool enough without any likelihood of distortion, the mould opens along the parting lines and the moulded object is ejected. The cycle is then repeated. Since injection moulding is a high temperature and pressure operation, the mould must be sturdy enough to withstand the extreme temperature and pressure.

Advantages:

a) Production rate can be very high due to the use of multi-cavity moulds.
b) It can handle a wide range of materials.
c) Excellent dimensional control is possible and scrap level is generally low.

Disadvantages

a) High investment in equipment and moulds is a must.
b) As a consequent of high cost, it is not suitable for low volume production.

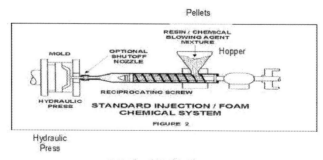

Injection Moulding Process

Figure 6.1

BLOW MOULDING

Many plastic bottles, particularly those with narrow openings, are moulded by one of the blow-moulding processes. There are many variations of blow moulding operations, but the three most prominent will be discussed. No matter the variation, all blow-moulding processes have the following sequence of operations in common:

a) A hollow plastic tube or parison is extruded.
b) While it is hot and in mouldable state, the parison is captured between two matched halves of a mould.
c) Air under pressure is introduced into the hollow parison, stretching the parison to conform to the mould wall or surface.
d) The formed bottle is held in the mould until it cools sufficiently to retain its shape. The mould is then opened and the bottle is removed for the next moulding cycle to commence.

The three variants to be considered are:

I. INJECTION BLOW MOULDING (IBM)

For this variant, the stages for container moulding are:

a) Parison is injection moulded around a blowing rod.
b) Injection mould is opened and parison is transferred into a blow mould.
c) Molten polymer is inflated with high-pressure air to conform to the blow mould.
d) When the formed container is sufficiently cooled, the blow mould is opened. Then, the container is removed and the cycle is repeated.

Polyethylene, Polypropylene, and Polystyrene are the commonly used polymers in injection blow moulding process. The process is suitable for smaller containers with no handles, such as the cosmetic jars, bottles and vials.

Advantages

i. Polymer distribution is uniform.
ii. Generally, there is no need for trimming because there are virtually no flashes.

Disadvantages

A given set gramme weight has to be adhered to. This cannot be changed unless a new set of blow moulds is built.

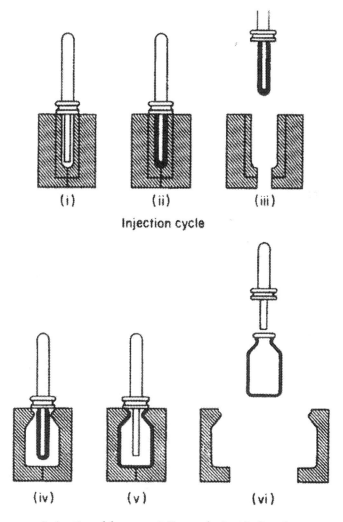

Injection cycle

Injection blow molding of plastic bottles.

Source: Plastics in Packaging Conversion Process, Institute of Packaging (1983)

Figure 6.2

II. EXTRUSION BLOW MOULDING (EBM)

The moulding steps are as follows:

a) Parison is extruded.

b) The extruded parison is pinched at the top and sealed at the bottom around a metal blow pin as the two halves of the mould come together.

c) The parison is inflated so that it takes the shape of the mould cavity.

d) When the formed container is sufficiently cooled, the mould is opened to remove the solidified container and the cycle is repeated.

The polymers generally used are the polyethylene, polypropylene, and polyvinyl chloride. The extrusion blow-moulding process can handle a wide variety of container shapes and neck openings, such as bottles with or without integral handles, jars, cans, and drums.

Advantages

Extrusion blow moulds are generally much less expensive than injection blow moulds and take much less time to fabricate.

Disadvantages

a) High scrap rate is generated.

b) There is limited control over the container wall thickness.

c) There is some difficulty in trimming the flashes (excess polymer).

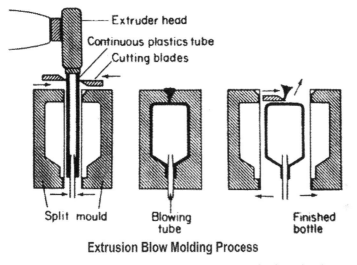

Extruder head
Continuous plastics tube
Cutting blades

Split mould   Blowing   Finished
               tube   bottle

Extrusion Blow Molding Process

Source: Plastics in Packaging Conversion Process, Institute of Packaging (1983)

Figure 6.3

III. INJECTION STRETCH BLOW MOULDING (ISBM)

There are two variants of this, namely the one-stage process and the two-stage process. In the one-stage process, the pre-form production, stretching and blowing all take place in one machine. However, in the two-stage process, the pre-form is produced in one machine, while the stretching and the blowing take place in another machine, the Re-Heat Blow machine (RHB). Since the two-stage process is more common, for various reasons, the moulding steps that follow are based on it. The moulding stages are as follows:

a) Pre-forms are produced by the injection moulding process and the neck finish of the bottle is formed at this stage.

b) The pre-form is heat conditioned and then transferred to the stretch blow mould. Here, the pre-form is mechanically stretched in the vertical direction while, at the same time, it is being stretched by air in the horizontal direction and a high pressure air inflates the pre-form to expand the bottle to its final shape.

The main polymer used predominantly in Injection Stretch Blow Moulding is PET, though some polypropylene and polyvinyl chloride are also used. The two-stage ISBM process is best suitable for high volume items and for both the wide mouth containers like the peanut butter jars, narrow mouth water bottles, liquor bottles, and carbonated beverage bottles.

Advantages

The biaxial stretching of the material increases the tensile strength, barrier properties; drop impact, stiffness, clarity, and the top load characteristics of the container.

Disadvantage

The only major disadvantage is the high cost of equipment and moulds, and the sourcing of the pre-forms.

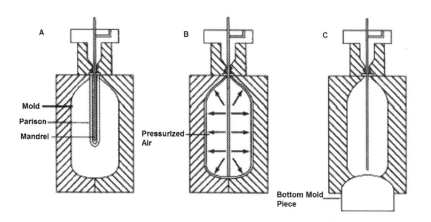

Injection Stretch Blow Moulding Process

Figure 6.4

THERMOFORMING

In this process, a plastic sheet is softened by heat and then forced either into or over a mould. There are three main methods of thermoforming, namely vacuum forming, pressure forming, and forming between matched moulds. Nowadays, there are many variations of the basic types of thermoforming, with one improvement goal or the other in mind. The goal could be to improve on processing, to have better control on wall thickness, or to be able to handle a hitherto more difficult material, etc. The three main methods of forming will now be discussed.

i. **Vacuum Forming Method**

The simplest vacuum forming has the plastic sheet clamped over a box containing the mould. Electric panel heaters heat the sheet. The air in the box is withdrawn through holes in the mould, thus creating a vacuum between the sheet and the mould. The atmospheric pressure of the air on the sheet forces it unto the mould, where it cools sufficiently to retain the shape when removed from the mould.

ii. **Pressure Forming Method**

This is similar to the vacuum forming method, except that air pressure is applied from above to push the softened sheet onto the mould. One important difference is that the pressure that can be applied during forming can be greater than the atmospheric pressure, so that better mould definition can be obtained.

iii. **Matched Mould Forming Method**

In this process, the heated sheet is pressed into shape by trapping it between matched male and female moulds. The method is particularly suitable when complex shapes are to be formed.

In all the three methods, heating of the sheet is normally carried out using infrared radiant panel heating. With materials such as polyethylene, polypropylene, and expanded polystyrene, double sided heating is usually required because of their combination of low thermal conductivity and high specific heat.

The major polymers or materials used in thermoforming are the polystyrenes (expanded, general purpose, high impact, etc), polypropylene, polyvinyl chloride, and amorphous PET.

Areas of application of thermoformed wares are thin disposable containers such as trays, tubs and lids, and tapered cups.

Other areas of application are in the packaging of electronic items and spare parts where they form trays/boxes in carded packages, and in medicinal tablets/capsules as blister packs. Also, trays, tubs, and lidded cups so produced are widely used for dairy products, such as soft cheeses, yoghurts, ice cream, and desserts.

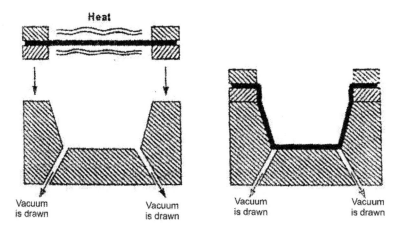

Simple Vacuum Forming Process

Source: Fundamentals of Packaging Technology, by Walter Soroka, Institute of Packaging (1996)

Figure 6.5

Advantages

1) It is a very economical way of making plastic packages, which are generally thin walled. It is a very good way of source reduction.
2) Thermoforming is suitable for low volume production runs.
3) Moulding pressures are low and moulds are relatively cheap.
4) Production rates are high.

Disadvantages

1) There are limitations on the types of shapes that can be molded, compared to other moulding methods.
2) The polymer must of necessity be formed into a sheet before thermoforming can take place.
3) Scraps are generally high.

Flexible Packaging Materials And Laminates

These are the materials that can easily be produced to very thin gauges for wrapping and labelling of goods either manually, on semi- or fully automatic machines. Incidentally, there is no universal agreement on the range of gauges normally encountered in flexible packaging materials. However, one basic fact is that all flexible packaging materials are capable of being bent or folded into whatever form or shape, even in the cold state. Flexible packaging is the most versatile sector of the packaging industry. It varies from simple single layer films, such as LDPE films through to two or three layer laminates and co-extruded films to multi-layer laminates with PET film, metallised film, aluminium foil, PVDC, and paper as layers.

The most common materials used in the flexible packaging industry are paper, various plastic films such as LDPE, LLDPE, and HDPE, metallocene, oriented and cast PP, regenerated cellulose, PET, PVC, Nylon, and surlyn. Material modifications and combining two or more of the materials into laminates constitute the backbone of the flexible packaging industry. When modifying the individual materials or combining two or more into a laminate, the goal the end product is meant to achieve must be in focus. The end result of such exercise constitutes unlimited options and packaging specifications for a diverse range of products. A run through the various materials that constitute flexible packaging will reveal their strengths and weaknesses.

I. PAPER

This is essentially a wood-based product, having been made from cellulose fibre derived from wood. Ordinarily, paper is not very functional as a packaging material because of its porous nature and its non-resistance to water, grease, oil, and fats. But with various treatments for different purposes, or with combination with other flexible materials, its functional performance becomes highly enhanced. Some of the treatments that enhance its performance and functionality are as follows:

a. Incorporation of special treatment during papermaking, such as clay coating, starches, and resins.
b. Coating with wax or any plastic material to make it water, grease and oil resistance, and to reduce its porosity.
c. Lamination to other plastic materials such as plastic films and aluminium foil improves its barrier properties.

II. REGENERATED CELLULOSE (CELLOPHANE)

Regenerated cellulose is made of pulp, which is specially treated to form a clear transparent film. It can be beautifully printed on a flexographic press or on a rotogravure press. It runs smoothly on automatic packing machines. It only becomes heat sealable when it is coated with polyvinylidene chloride (PVDC). This coating also enhances its barrier properties.

Regenerated cellulose is used all over the world to wrap, twist-wrap, or over-wrap sweets and other products. However, in recent times, other films such as PVC and cast PP have overtaken regenerated cellulose in these roles due mostly to their cost advantage over cellophane and their availability.

III. PLASTIC FILMS

Plastic films are the most widely used flexible packaging materials. In volume and diversity, they have no rival. The major plastic materials commonly used are the polyethylenes (PE), polypropylene (PP), polyvinyl chloride (PVC), and polyethylene terephthalate (PET). There are many grades and variants of each of these materials. Most of the plastic films are made by the extrusion process.

Plastic Film Extrusion

There are two main methods of manufacturing plastic films, namely:

1. Cast film (slit-die) extrusion
2. Blow extrusion

Both start with the extrusion of molten polymer. The difference only lies in the design of the die through which the polymer passes and in the subsequent haul-off. Since the slit-die, cast film extrusion method is usually employed for sheet manufacturing, the description below will be based on the Blow extrusion process, which is the method normally used for manufacturing films.

Blow Extrusion

In the blow extrusion, the plastic resins are fed through a hopper into the extruder. The granules are carried along the extruder barrel by the rotation of the screw. As they progress, they are melted by the contact with the heated barrel and by the generation of frictional heat. The molten plastic is forced through the die, but just before this point, it passes through a screen pack, which consists of a number of wire meshes. The function of the screen pack is two-fold. First, it acts as a filter for any contamination that may be present in the raw material. Secondly, it tends to restrict the flow of the molten resin and so creates a backpressure. This backpressure improves the mixing and homogenisation of the molten plastic and increases the frictional heat. As the molten plastic enters the die, it is made to flow around a mandrel where it emerges from a ring-shaped die opening in tubular form. The tube is blown into a bubble by air introduced via the mandrel. The air is trapped in the bubble by the die at one end and the pinch rolls at the other. The size of the bubble and the thickness of the film are controlled by the extrusion speed, the haul-off speed, the die-ring width, and the air pressure within the bubble. The resultant collapsed tubular film is known as lay flat film and can be used as such for bag or sack making or it can be slit and made into flat film. Blow film extrusion process is used to manufacture low-density polyethylene, high-density polyethylene, and some polypropylene films. The technique is used to produce tubing, flat, or gusseted films.

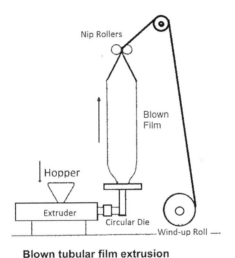

Blown tubular film extrusion

Figure 7.1

Low Density Polyethylene (LDPE)

This material is heat sealable. It is oil, grease, and water resistant. It can be printed on both flexographic and gravure presses. It has poor clarity; that is, it is translucent. It is the major material used for shopping or carrier bags. It is used on form/fill/seal machines to form bags or sachets, which are filled with products before sealing. It can also be processed for use as shrink and stretch films.

High Density Polyethylene (HDPE)

Most attributes of HDPE are similar but superior in all respects to that of LDPE. It is tougher, has better barrier properties, and can withstand higher temperature environment than LDPE. Uses of HDPE film are essentially the same as LDPE film. It is rather milky in appearance and tends towards being opaque.

Polypropylene (PP)

PP has a lower density than LDPE and HDPE and hence, gives higher yields gauge for gauge. It has higher barrier properties and greater stiffness than the polyethylenes. It can also withstand higher temperature. Uses include those mentioned above for the polyethylenes, but, additionally, it is used for wrapping snack foods and chocolate bars where the material has been modified or processed into clear, opaque white, pearlised or metallised film in each case to achieve some end use goal.

Polyvinyl Chloride (PVC)

PVC film is transparent. On ageing and with exposure to sunlight, it tends to turn yellow. Hence, in order to mask the yellow color, a blue colorant or masterbatch is added during processing. It has good gas barrier property, but the water vapour permeability is high, making it unsuitable for wrapping highly hygroscopic products. It is used generally as stretch and shrink films and also in twist wrapping of sweets and candies. The disposal of this material by incineration has been banned in many countries because of its adverse effect on the environment.

Polyethylene Terephthalate (PET)

PET film has excellent clarity. It is very strong and therefore can be used at low gauges. The film can be used in plain form and can be reverse printed and laminated to other flexible materials such as aluminium foil or other plastic films. It can also be metallised to enhance presentation and/or improve barrier properties. Without metallisation or lamination to other materials, its barrier properties are just moderate. PET film is a common outer component or layer in most laminates in which aluminium foil is a barrier layer and where reverse printing is desirable. The excellent clarity makes it a clear choice in such laminates. Such foil incorporated laminates are very suitable for oxygen and water vapour sensitive products such as cocoa-based beverages, powdered milk, and other hygroscopic products.

Aluminium Foil

This is a rolled section of pure aluminium to a very low gauge. When it is intended to be used as a layer in a laminate, the gauge can be as low as seven microns. Nine microns is typical in most laminate structures. Because of its jelly-like nature, aluminium foil is rarely used unsupported unless at fairly high gauges.

Advantages of aluminium foil are:

a) It has excellent barrier properties at nine microns and above.
b) It is non-toxic and inert.
c) It has good dead-fold characteristics.
d) It has very attractive appearance.
e) It can be printed on using flexography or gravure press.

Disadvantages of aluminium foil are:

a. It is very costly.
b. It has low stiffness (almost jelly-like at low gauges) and hence needs to be supported by lamination to other materials.
c. It is not heat sealable and therefore always needs a heat sealable layer or coating in situations where sealing is necessary.

Flexible Packaging Material Modification

Many flexible packaging materials, when used alone (singly), will hardly deliver as far as functional performance is concerned. Therefore, in order to enhance their functional performance, they are subjected to various modifications. The major modifications are described briefly below.

a. **Coating**

Due to the high porosity of paper, it is wax or ethylene vinyl acetate (EVA) or polyethylene coated to improve its barrier properties and to make it grease, fat, oil and water resistant. Also, regenerated cellulose (cellophane) is PVDC coated to make it heat sealable and to improve its barrier properties.

b. **Co-Extrusion**

In many packaging applications, single-layer films are hardly suitable to meet all functional requirements, hence the need to marry or combine two or more layers of film of same or different materials. One of the ways of doing this is by a process called co-extrusion. In co-extrusion, two or more extruders are coupled to a single common die to produce two or more layers of plastic film, which has superior properties to any of the individual layers.

Advantages of co-extruded films are:

i.   The cost is lower compared to laminated constructions.
ii. It has lower tendency to delaminate.
iii. There is greater flexibility in obtaining a wide range of performance characteristics.

Disadvantages of co-extruded film:

i.   It is difficult to utilize the scrap due to strong bond between the different layers of material.
ii. Sandwich or reverse printing is not possible.

c. **Metallization**

Film metallization is a process in which a metallic coating (e.g. aluminium) is deposited onto a plastic film or paper by vaporising the metal in a high vacuum chamber and allowing the metal vapour to condense onto the slow-moving substrate web. It results in the most economic use of aluminium.

Advantages of metallization are:

i. It provides the aesthetics look of foil without the high expense.
ii. It provides reasonable barriers against moisture vapour and oxygen. The level of protection depends on the coating weight.

d. **Lamination**

This is the backbone of the flexible packaging industry. Realising the apparent shortcomings in most single layer flexible packaging materials, the need arose to combine two or more of these materials in order to achieve the desired goals. Therefore, lamination is a process of combining two or more dissimilar flexible materials to form a multilayer material called "Laminate". The idea is to tap into the area of strength of each component to obtain some sort of synergy in material usage/functionality to provide better and superior product protection and preservation.

The major lamination methods are:

i. Wet bonding (water based adhesives) lamination
ii. Dry bonding (solvent based adhesives) lamination
iii. Hot melt (solventless) lamination
iv. Extrusion lamination
v. Thermal (wax) lamination

No attempt will be made to go into the detailed processes of the lamination methods, nor will the advantages and disadvantages of each method be discussed because it is beyond the scope of this book.

Uses of Flexible Packaging Materials

Flexibility in use is wide and pervades the whole range of industries, from foods, drinks and pharmaceuticals, to non-foods. The food industry, particularly the snacks, confectionery and fast foods, constitutes the highest users of flexible packaging materials. The forms in which these materials are used also vary widely. The major forms are:

   i.   Labels
   ii. Liners for boxes, drums, etc
   iii. Wrappers
   iv. Over-Wrappers
   v.   Bags
   vi. Envelopes
   vii. Pouches: pillow pouches, 3-side seal pouches, and 4-side seal pouches.

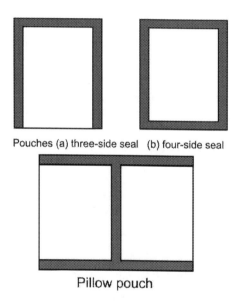

Pouches (a) three-side seal   (b) four-side seal

Pillow pouch

Pouches
Figure 7.2

What Makes The Flexible Packaging Industry Tick?

a) It is cost-effective due to high yield, which brings down the cost per unit pack.
b) It is readily available in a wide variety of forms, from simple single (mono) layers to complex multilayer laminates.
c) Origination costs vary from low (one-colour) flexographic printing to very expensive, six to eight colour gravure printing.

All of the above make flexible packaging one area of the packaging industry where the low profile MSMEs as well as the high-profile multinational giants can find accommodation. It is that versatile and flexible.

Packaging Printing
And Decoration

Three elements define a package. They are the material, form (structure), and design (graphics). Any type of packaging material can be manipulated to give us any form or shape, while the resulting shape can be printed or decorated to any level of sophistication with the appropriate graphic design. The end result of the third element, graphic design, which defines a package, is the subject of this chapter.

Why do we print packaging materials?

i.   It is a means of branding or product identification.

ii. It enhances the aesthetics of packaged goods and therefore stimulates product sales.

iii. It is a medium of communicating important information to the consumers or users of the products.

iv. It is a means of meeting government requirements concerning different categories of products.

v.   In our globalised world of today, packaging has largely become the salesman, albeit a silent one. Hence its relevance to a product's success cannot be over-emphasised. See a write-up on this at the end of this chapter.

Most printing processes are essentially the selective application of ink or paint to the packaging material (substrate) from a cylinder, plate, stereo, or squeegee by direct or indirect contact with the packaging material to be printed. Without printing, consumer goods packages will be so dull and drab that the consumer will have a lot of difficulty in distinguishing and choosing between product brands on display. The role of colour and typography in package printing is so crucial that, without it, the success of communicating the relevant information about a product to the consumer will be very limited. Before the emergence of the computer in the graphic design trade, artwork preparation used to be a tedious and painstaking assignment. Not only was getting the right colour a trial and error job, typography was also a specialized trade before the digital age. Today, the computer takes care of all of these in a fast and efficient manner.

From the three primary additive colours, red, green, and blue are generated the three primary process colours (yellow, magenta, and cyan) used in colour printing.

> Red + Green = Yellow
> Red + Blue = Magenta
> Blue + Green = Cyan

All other colours are produced by manipulating these three process colours. The black colour is used as key colour to add depth where and when necessary. The computer is now capable of generating a whole colour gamut that can be viewed, edited, displayed, and printed with appropriate typefaces to arrive at beautiful and eye-catching designs for the packaging materials of choice.

INK TRANSFER PRINCIPLES

There are several ways of transferring ink or other marking media onto the substrate or the material to be printed. However, about four major principles account for over 95 percent of package printing. The four principles are:

a. Relief

In the relief principle, the actual printing areas of the printing unit (e.g., stereo) stand out above the rest. Thus, when ink is applied to the unit by means of a roller, only the printing area comes in contact with the ink supply. It then follows that when the printing unit is brought into contact with the packaging material to be printed, only the printing areas are in contact, and ink is transferred or deposited on those areas.

RELIEF

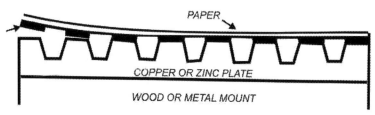

Source: National Extension College & Institute of Packaging, UK (1978)

Wood Or Metal Mount
Figure 8.1 Relief

b. Intaglio

This is the exact opposite of the relief principle, for the printing areas are recessed slightly below the general surface. Ink is usually applied to the whole surface, but is then scraped off from the non-printing area by a doctor blade, thus leaving the recessed areas as ink carriers.

INTAGLIO

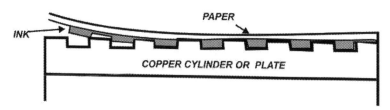

Source: National Extension College & Institute of Packaging, UK (1978)

Intaglio
Figure 8.2

c. **Planographic**

This is a compromise (or a hybrid) of the relief and the intaglio principles in that both the printing and the non-printing areas are at the same level and both are contacted by the printing roller. The non-printing area must therefore be treated in some way so that the ink is repelled there.

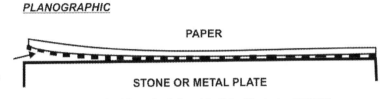

Planographic

Figure 8.3

d. **Stencil**

Here a screen is treated, either mechanically or photographically, so that the non-printing areas are impervious to ink.

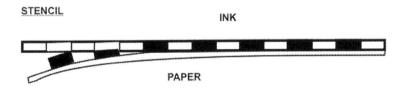

Stencil

Figure 8.4

Other printing (or ink transfer) principles that are now emerging in the scene and are already in use include inkjet printing, heat-transfer printing (hot foil), embossing, etc.

Each of the conventional printing processes that follow adopts one of the four ink transfer principles in achieving its printing goal.

The Major Printing Processes

The major printing processes are Letterpress (and variants of it), Flexography, Lithography, Rotogravure, and Screen printing.

The four that are very popular with consumer good packaging printing are the flexography, lithography, rotogravure, and screen printing processes and these four will now be described below.

Flexography

This is based on the relief principle. Flexographic process is widely used for printing plastic films, cellophane film, aluminium foils, and corrugated fibreboard cases. For the non-absorbent films, ink is solvent-based and drying is largely by evaporation. Design is engraved on rubber stereo but lately on photo-polymer stereo, which is wound round the plate cylinder. Excess ink is applied to the engraved transfer/anilox roll and the transfer roll meters the correct amount of ink to the raised surfaces of the printing plate attached to the plate cylinder. The packaging material is passed between the plate cylinder and the impression cylinder to achieve ink transfer. Ink drying takes place between printing stations to avoid smudging. The above sequence of events is repeated for each ink color. See Figure 8.5

Advantages of Flexography Printing

i.   It is suitable for printing fine and rough surfaces.
ii. A wide range of colours is possible.
iii. Low and affordable stereo cost is an attraction to most MSME operators.
iv. Ink drying is rapid (by evaporation) and hence enhances productivity.

v. Changeover is quick and this also enhances productivity.

vi. It is suitable for small, medium, and high volume jobs.

vii. The print quality continues to improve, but certainly not as good as gravure's.

Disadvantages of Flexography printing

i. It is sensitive to changes in printing pressure. Pressure variations appear as color variations in finished jobs.

ii. The image areas tend to "squash" when they contact the substrate.

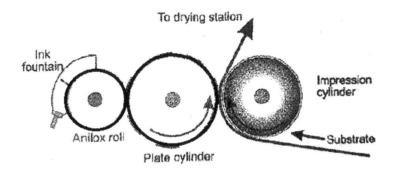

Source: ITC Packit - Packaging Design (2005)

Flexographic Printing Process
Figure 8.5

II. Lithography

Lithographic printing process is based on the planographic principle of ink transfer. In this process, both the printing and non-printing areas are at the same level relative to the inking roller, and both make contact with it. Therefore, the plate has to be treated so that it attracts water and repels ink in the non-printing areas, and do the reverse in the printing areas—that is, attracts ink and repels water.

A rotary offset lithographic machine consists essentially of ink metering system, a dampening system, a material feeding system, a metal plate on a cylinder; a rubber offset blanket cylinder, and an impression cylinder with a resilient surface to ensure good contact with the material to be printed. See Figure 8.6.

The plate cylinder rotates first in contact with a dampening roller; next, in contact with the inking rollers, where the oil based ink is repelled by the water film but accepted by the printing (image) areas. At the first nip between the plate and the blanket cylinders, ink is transferred (offset) from the plate to the blanket. This is in turn transferred to the material to be printed when the latter passes between the blanket and the impression cylinders. Lithography ink is oil-based and, consequently, drying takes place largely by oxidation. Lithography is mostly used for the printing of all types and styles of cartons and labels. It is also the main process used for the printing of tinplates.

Advantages of Lithography

i. It prints clear, sharp, and well-defined images.
ii. Printing plates are inexpensive and can be easily made. Plate replacement is cheap.
iii. It is economical for small runs.
iv. It prints well on metal surfaces and on smooth carton-board surfaces.

Disadvantages of Lithography

i. Oil-based inks dry very slowly and drying may need to be aided by drying powder.
ii. Sheet-fed lithography is slower than web-fed flexography or gravure.

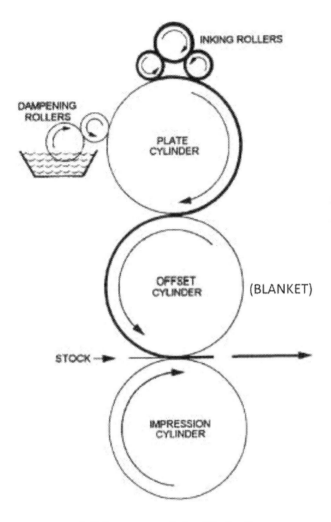

Lithographic Printing

Source: ITC Packit - Packaging Design (2005)

Figure 8.6

III. Gravure (or Rotogravure)

This is an intaglio-based process. Rotogravure is the ultimate in package printing, with an unbeatable level of sophistication and quality. The printing cylinder is made of copper-plated mild steel, which is polished to smooth consistency. Etching or engraving of designs on cylinders used to be manual, slow and painstaking, and could take weeks or months. But with the advancement of technology and the digital age, etching of the design is now done electronically and this involves creating millions of tiny cells or wells that carry the ink. A cylinder is needed for each colour to be printed. The engraved cylinder rotates in an ink bath or reservoir. The excess ink on the surface of the cylinder is removed by a "doctor blade". The ink now left in the cells, the recessed image areas, is then transferred to the material to be printed. Gravure ink is solvent-based for easy drying and drying is by evaporation.

Gravure printing process is used for a wide range of materials in the flexible packaging area. The material to be printed is roll-fed into the printing system and the material surface must be smooth.

Examples of materials normally printed on the gravure press are plastic films like PET, PE, PP, paper and paper labels, laminates, and metallised films.

Advantages of Rotogravure

i. Excellent print quality with good colour illustration.
ii. It can handle complex multicolour designs.
iii. It is very fast and most suitable for high volume printing jobs. It is a process that only thrives on volume.
iv. It handles halftone jobs better than any other printing process.

Disadvantages of Rotogravure

i. It is very uneconomical for short runs; it is only suitable for long runs.
ii. Origination costs such as cylinder costs are very high.
iii. Cost of printing equipment is also very high.
iv. Cylinder making process used to be very long, but has now been substantially reduced with the use of electronic engraving.

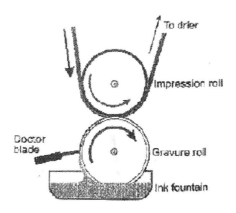

Gravure Station

Source: ITC Packit - Packaging Design (2005)

Figure 8.7

Screen Printing

This is a stencil process that was originally based on a silk screen. However, today the screen is most likely to be made of fine wire or nylon or polyester mesh. The screens usually have about 70-120 mesh counts (perforations) per inch (this varies a lot), and they are prepared photographically, being porous only in the areas where decoration is required. The screen is supported by a metal or wooden frame that holds the ink supply and keeps the screen taut.

As the material to be printed is positioned under the screen, thick pasty ink is spread on the screen. A rubber squeegee is drawn across the screen, thus forcing the ink through the porous areas of the screen onto the surface to be printed. The frame is raised and the printed article is taken out for hot air drying.

Screen printing is used mostly for printing irregularly shaped packages, such as plastic bottles, glass bottles, as well as flat items like textile materials, T-shirts, face-caps, towels, mugs and drinking glasses, jerry cans and plastic kegs, promotional items like key holders, handkerchiefs, etc.

Advantages of Screen Printing

i. Origination cost, the screen, is very cheap and can be easily prepared.
ii. It can print virtually any material.
iii. It is able to print with the largest variety of inks.
iv. Cost wise, it is the most appropriate process for the MSME operators with low capital base.

Disadvantages of Screen Printing

i. It can print only one colour at a pass.
ii. Production speed is low compared to other printing processes.
iii. Heavy ink lay down increases ink consumption and cost.
iv. Print quality is not top grade.

Factors that Determine the Choice of a Printing Process

i. The quality and characteristics of the material to be printed, whether porous or non-porous, smooth or rough, etc. is a factor.
ii. The quality of print required
iii. The quantity to be printed—that is, the length of run
iv. The form in which the printed material is required
v. The capital outlay
vi. Available facility or technology
vii. Any special requirements such as rub resistance, freedom from odour, etc.

SUMMARY OF PRINTING PRINCIPLES AND PROCESSES

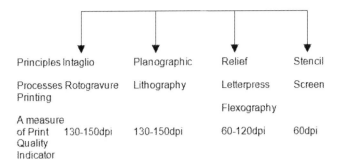

	Intaglio	Planographic	Relief	Stencil
Principles				
Processes	Rotogravure Printing	Lithography	Letterpress Flexography	Screen
A measure of Print Quality Indicator	130-150dpi	130-150dpi	60-120dpi	60dpi

Figure 8.8

Packaging - the Silent Salesman

Until recently, particularly in the developing nations, packaging has been considered a minor element in the marketing mix. The traditional objectives of packaging have been product protection and consumer convenience. Recent developments in retail practice have conferred some important status on packaging - that of sales appeal. The sales appeal aspect of packaging is so important in modern retail selling for so many reasons that a lot of attention has to be devoted to the surface design of packs. This is necessary because in many cases, it is the surface design of a pack that first appeals to consumers rather than the product itself, which of course is alien to the first time buyers. In such situations, the package surface design makes a lot of difference between selling and no selling.

In addition to sales appeal, the pack has to carry some identifying statements voluntarily or involuntarily without which the pack does not make any sense to the prospective buyers. Such information or identifying statements include:

1. the brand name of the product and its producer/packer. The primary role of branding is to distinctively identify the product, particularly among several similar competing brands in the same market.
2. A declaration of product ingredients, particularly for edible products on which also preservatives, colourants, etc, must be declared for public health and safety reasons.
3. Nutritional claims on vitamins, minerals and other essential ingredients on the basis of RDA are some of the information the packaging silently conveys to the consumer.
4. Direction for use - This is particularly essential in respect of medicinal and poisonous products.
5. Package average or minimum packed weight or filled volume is required to be declared on the pack where the Weights and Measures Acts are in force.

6. Other pieces of information the package carries are the manufacturing date, the batch number for traceability in the event of quality problems and the Best Before date to guide consumers.

7. In some sophisticated economies, prices are now printed on the individual packs for the convenience of shoppers at supermarkets and other self-service stores. In the alternative, bar code which carries a lot of information on the product is incorporated into the pack design. Scanning machines for decoding such information are now part of the facilities at the paying counters of most supermarkets worldwide.

All the above points about pack design have helped reduce the burden on the consumer considerably. For instance, consumers no longer need to call the attention of the storekeepers, as it used to be, before deciding whether or not to buy a product because all the relevant information required to make a buy or no buy decision are already on the pack. The presence of price tags or labels has also helped both the storekeeper and the buyer to overcome the problem of having to spend valuable time haggling and bargaining over the price to be paid for an item.

Considering the roles the pack design/labelling or identification plays in modern self-service retail outlets, it goes without saying that packaging is truly silently playing the role of the salesman.

Packaging Specifications

The authors' experience in the course of relating with several MSME operators is that many procure their packaging materials without any written down specifications. Some settled on verbal agreement with the packaging suppliers and there are instances where off-the-shelf packaging materials were just adopted, not for their suitability but only for their availability. This practice often results in the use of unsatisfactory packaging materials. The consequences of packing products in unsuitable packaging materials can be disastrous. No wonder packaging remains one of the major challenges facing the MSME operators in the developing countries. This also poses a major hindrance when it comes to exportation of their products to foreign countries.

First, what is a packaging specification? Secondly, why is it essential or mandatory to have it?

Definition: There is no known universal definition for a packaging specification. The definition we are providing here is borne out of the authors' several years of experience in the packaging profession. Thus, a packaging specification can be defined as an accurate and detailed description containing information about all the necessary facts, properties, and special features of a packaging material, to facilitate unambiguous communication between the manufacturer (supplier) and the user (buyer). When the specification has been agreed, it represents an agreement between the supplier and the buyer. This agreement must be adhered to until there is a mutual decision to alter or change parts of it or replace it completely. Specifications evolve as a result of development works, laboratory trials, field trials, and storage tests. Both the supplier and the buyer can be involved in these development works. By the time the specification is put in place, the buyer must have been convinced that the material is capable of meeting his product's requirements; while the supplier must be sure he has the facilities and the technical know-how to produce to the agreed specification. A packaging specification, therefore, is not an imposition of the buyer's wishes on the supplier and vice versa.

Words, pictures, drawings, and numbers are the building blocks needed to arrive at a suitable packaging specification and the specification writer's skill and experience must be used intelligently to arrive at a specification that is neither ambiguous nor controversial.

Why do we need Packaging Specification?

There are six major reasons (there could be more) why packaging specifications are mandatory, and these are:

i. To ensure a smooth packaging operation on production lines and to have a packaging that will sell the required product successfully in the marketplace.

ii. The specification represents the agreement (or bond) between the supplier and the buyer. Non-adherence to the contents of it constitutes a breach.

iii. To avoid misunderstandings regarding both the technical and commercial details in the transaction.

iv. To enable the buyer to search for alternative suppliers and facilitate a comparison of the different offers received.

v. To provide the user staff with a basis for accepting or rejecting incoming packages and components.

vi. To serve as a basis for settlement of eventual claims, if any should arise.

What information must a packaging specification contain and in what format is the information presented?

Essentially, the packaging material specification, in whatever format, must contain in details the following pieces of information:

Company Name: This is the name of the buyer/user company that owns the specification.

Packaging Item: A display box, wrapper, bottle, etc.

Reason for issuing the specification: Is it for a new product, a replacement for an existing material, or a review?

Reference Number: Format depends on individual companies.

Effective Date: Date of commencement of use of the specification.

Scope: This is a brief description of what the packaging material is supposed to do.

Material: Here, the material is described in details, e.g. 400gsm white line chipboard with +/- 5 percent tolerance on the basis weight.

Construction or Style: Full description of style with basic dimensions and tolerances on the dimensions. A detailed technical drawing of the material blank for a unit package is a must for full understanding of what is expected.

Printing: Here, the artwork and possibly the colour separations, colour standard, tolerance levels with other expectations, such as scuff resistance, ink stability to UV light, etc. must be presented to the printer/converter of the packaging material.

Packing/Delivery: Depending on the type of packaging material, packing and mode of delivery must be clearly spelt out here.

Classification of Defects: Usually, three defect levels are recognized worldwide and these are Critical, Major, and Minor. Which defects fall into each level must be clearly and carefully identified and the tolerance level (percent allowed) for each level agreed.

Preparing or writing a packaging specification requires a full understanding of which performance factors are critical. Is it product protection, smooth running on the machine, or ability to withstand warehouse stacking and distribution hazards, and many more?

Nowadays, with the emergence of computer and many software packages, all of the above can be generated and made available to a supplier or printer online for flawless execution.

Below is a typical packaging material specification for a folding carton meant for packing 200gm of a detergent powder.

Specification for a Folding Carton for 200gm Detergent Powder

Company Name: ABC Company Ltd

Packaging Item: Display carton for 200gm of Detergent Powder

Reference Number: X X X X X

Effective Date: August 15, 2012

Scope: This specification describes the display carton or box to contain 200gm of a detergent powder produced for nationwide distribution.

Material: The material of construction shall be 350 g/m^2 (gsm) white lined chipboard (WLC).

Construction: Blanks for the carton shall be dimensioned as per the specification drawing number YYYYY. The grain direction of the board (blank) shall be horizontal to the print as shown on the specification drawing. The board substance or grammage tolerance shall be +/- 5 percent, while the tolerance shall be +/- 0.5 mm on any of the basic dimensions.

Basic Dimensions (Internal): Length = 118 mm; Width = 36 mm; Height = 180 mm.

Printing: This shall be by offset lithography or by the gravure process. Printing must comply with the agreed artwork and colour guides. The print must be light stable while the varnish must be scuff resistant—that is, the print must not lose colour or fade on prolonged exposure to the UV light, nor must it abrade when rubbed against itself or any similar surface, plain or printed. Print must be stable to water, soap jelly, and 5 percent alkaline solution. Finally, the inks as well as the board on which the printing is done must be odour-free.

Gluing: There must be evidence of fibre tear on every glued area when pulled apart. There must be no under-gluing, which may result in glue failure or over-gluing, which may make carton erection on the packing/cartoning machine difficult or impossible. The glue must be odour-free.

Packing and Delivery: Cartons in flat or collapsible form must be wrapped up in bundles of fifty or any convenient number and twenty or so of such bundles be arranged edge-on in corrugated boxes or in lined wooden boxes. No visible damage must be evident on any part of the cartons supplied as described above. Every consignment must be protected in transit against any inclement weather, such as rain or dust as wet and/or dusty cartons will be rejected. A quality report must accompany every consignment.

Inspection: Regardless of any accompanying quality report or certificate of analysis, the normal entrance control checks will still be carried out on every consignment as outlined elsewhere in the receiving company's QC Procedures Handbook. Any detected defects that are over the tolerance levels specified will be rejected if a repeat sampling confirms a higher level of defects above the tolerance levels as indicated below.

Classification of Defects

Critical or Class A Defects: These are defects that make the cartons unusable or, if used, may pose a threat to the survival of the product. Examples of such defects are tears, holes, glue failure, odour, wetness, wrong grain direction (if cartons are to be run on automatic machine), asymmetric layout, etc.

Major or Class B Defects: These are defects that may result in marginal or borderline functionality or low quality appearance. Examples are low grammage, small deviations in dimensions, color variation, smudged printing, poor crease quality, poor stability to UV light, etc.

Minor or Class C Defects: These are minor faults affecting mostly appearance. Examples are slight off-colour, rough printed surface or low gloss, improper packing for delivery, etc.

Maximum Allowable Defect Levels

Class A or Critical --------------------------------0.5 percent

Class B or Major ------------------------------2.0 percent

Class C or Minor ----------------------------5.0 percent

Prepared by: ---------------------------------   Date:

Checked by: ---------------------------------   Date:

Approved by: ---------------------------------   Date:

Causes of Packaging Failure

There are several reasons why packaging failure could happen. The following come readily to mind:

1) Wrong choice of packaging material could lead to a serious failure, particularly if the packaging material cannot adequately protect and preserve the product.
2) Delay in initiating a packaging development work for a product can lead to a failure. Not allowing enough time for development

could result in hasty and shoddy job lacking in well thought out development program.

3) Failure to contact packaging experts, particularly when the relevant expertise is not available in an organization, is a recipe for failure.

4) Failure to carry out market research and test marketing, when we are not absolutely sure that all relevant parameters have been adequately taken care of, can lead to failure.

5) Doing product and packaging development separately can be a costly mistake. Product development experts and their packaging counterparts must work together right from the product concept stage through development stage to full implementation.

6) The use of materials not complying to agreed specification could lead to packaging failure.

Consequences of Packaging Failure

The consequences of packaging failure are many and the following are the obvious ones:

1) It is always a shock and a great embarrassment to any corporate body to learn about a packaging and or product failure.

2) It is a huge cost to companies to carry out inspection and investigation through field trips to unravel the cause or causes of failure.

3) For less serious cases of failure and where the consequences of it do not pose any threat to the consumer, discount can be offered to the consumer as an inducement for them to buy off the product as soon as possible.

4) For very serious cases where product recall is inevitable, the cost of organizing recall, transporting the goods from the various outlets across the country can be very prohibitive.

5) If the recalled goods have to be reprocessed, the cost of doing so is usually more than the cost of producing it from scratch. The reprocessing could mean a total loss of packaging materials.

Handling of recalled products from the shop floor could be very messy and demoralizing to the work force.

6) If the situation is so bad that a total write-off and subsequent destruction is inevitable, the total cost of producing the goods and all the attendant efforts would have been wasted. Superimposed on this will be the cost of destruction and the disposal of the resulting waste.

7) Failure of whatever nature saps staff morale and sometimes creates internal friction resulting from buck-passing, if not carefully managed.

8) Finally, the biggest loss is that of customer goodwill, which is not easily quantifiable. When failure results in product withdrawal or recall from the trade, the memory of the episode lingers on in the mind of the consumer for a long time. It takes a lot of corporate efforts and resources to erase the negative image already created in the mind of the consumer.

If packaging failure could lead to all or most of the listed consequences, then the importance of consulting a packaging expert at the preliminary stage of product development cannot be over-emphasised.

Therefore, the following must be considered at the blueprint stage of packaging development.

1. Dynamic nature of packaging.
2. Understanding the consumer behaviour.
3. Every product requires its own unique packaging requirements.
4. Cost of packaging compared to product cost and its consequences on the selling price.
5. Convenience, ease of disposal, environmental issues.

Packaging And The Environment

In the last couple of decades, the environment has been a topical issue worldwide. This is so because our environment has become so bastardised and degraded that unless something is done urgently, human existence may become endangered sooner than later. The mess the environment has become is due largely to the uncontrolled activities of man. As a result of increased population and our extravagant and wasteful lifestyle, we distort the ecosystem by the amount of waste we dump into the environment. These wastes include all sorts of solid waste through our excessive demand for consumer and luxury goods, some of which are grossly over-packaged, greenhouse gas emissions from fuel guzzling automobiles, airplanes, factory fumes, and wasted electricity and water. Through these activities and others, we are making the environment less and less conducive for human living, even though we shout the need for environmental sustainability each time the world leaders meet to debate environmental issues.

How Does Packaging Come Into This?

All the consumer goods we buy on daily basis and most other goods are packaged for the various reasons already discussed in the previous chapters. At the point of consumption of these goods, their packaging materials have to be removed in order to access the product—be they food items, drinks, pharmaceutical products or machine/equipment accessories. When these packaging materials are disposed off irresponsibly, they constitute a litter and a nuisance to the environment. Therefore, packaging is a contributor to the solid waste problems that countries and cities all over the world have to contend with. Available information shows that in most of the developed countries, packaging materials account for about 33 percent of the municipal solid waste. We have no information on what the proportion is in the developing countries. We believe the proportion will be much less because packaged goods consumption in the developing countries, particularly food items, is much lower than what obtains in the advanced countries.

Should we, because of the solid waste mess, ban packaging? The answer is no for various obvious reasons. A ban on packaging will be like turning back the hand of civilization. It will lead us back to the era preceding the industrial revolution when most goods, particularly the agricultural produce, were seasonal and restricted to the local environment where they were grown with the attendant scarcity and famine. It will spell a doom to the international trade, which is driven by packaging. If we cannot ban packaging, then what is the way out? The way out is the proper and responsible management of solid waste. An item of whatever value becomes a litter *once it is placed in a wrong place*. Most of the challenges we face as far as solid waste management is concerned are man-made, absolutely self-inflicted. Must we throw empty bottles and bags, sweet or biscuit wrappers carelessly and indiscriminately on the streets and highways where they will be blown by the wind or carried by erosion into the drainage systems and ultimately block the drains to cause flooding? Such socially irresponsible behaviour only compounds our environmental problems.

In order to reduce the amount of used packaging materials that go into the waste stream, a number of initiatives have been taken and these are constantly being improved upon. These initiatives have now been christened as the four "Rs"; that is, Reduce, Reuse, Recycle, and Recovery.

Reduce

This is an initiative that aims at encouraging minimizing the amount of packaging material needed to take care of a product without compromising its quality. It is a source reduction scheme that tries to strike a balance between under packaging and over packaging, both of which have adverse consequences. For example, over the years, the weight of glass bottles has been reduced by as much as 25 percent through light-weighting without any reduction in strength.

Reuse

One other way of reducing the amount of packaging going into the waste stream is to design it in such a way that it can be used over and over again. One area where this initiative has been very successful is the soft drink and the beer industry, where glass bottles are being returned to the brewers for rewashing, refilling, and then sent back to the trade.

Recycle

This is another initiative that involves collecting materials that ordinarily would have ended in the waste stream; clean them up and reprocess them into useful products. In recycling, some of the end products could be as good as the initial product from which the recycled materials originated. A good example of this is the use of broken glass, "cullet", in glass making. The glass container produced using the cullet as part of the fresh feed into the furnace is not inferior to the one made without the cullet in the feed. In fact, the use of cullet in glass making substantially reduces energy requirements. On the other hand, the use of recycled materials could result in some product downgrading. An example is the case of using recycled plastic material in making shopping bags or refuse disposal bags. Whichever way, the objective of reducing the amount of packaging materials entering the waste stream is still achieved.

Recovery

This is a variant of the Recycle initiative. But it is fast acquiring a personality of its own—hence, we talk of 4Rs instead of 3Rs. Essentially, the goal is to extract the heat from burnt/incinerated wastes, particularly the ones with high calorific value like plastics, to generate steam for space heating or electric power and the use of hot combustion gases to power turbines to produce electricity. It is a waste to wealth (energy) scheme, which is already gaining popularity.

Having discussed the four major ways by which packaging materials going into the waste stream can be reduced, we now look at what is done with what inevitably goes into the waste stream.

Solid Waste Disposal Methods

There are three common methods of solid waste disposal, namely open dumps, sanitary landfills, and incineration, with some of them now having some variants.

1) **Open Dumps:** An open dump, as the name implies, is a place where refuse is dumped in a pile and left to moulder for all time. It is the most primitive form of solid waste disposal. Another form of open dump is "open dump with burning". Because of obvious health hazards, potential for accidental or deliberate fires and direct "contribution" to the greenhouse gas emissions, open dumps, with or without burning, are no longer encouraged anywhere.

2) **Sanitary Landfills:** Sanitary landfill is a carefully engineered project where refuse is carried to a pre-prepared site, compacted, and covered with a layer of soil each day. At the completion of the operation, the compacted refuse may be several feet deep, and the final layer is then covered with six to eight feet of soil. The terrain so created is suitable for the construction of parks, golf courts, and some light buildings. One big disadvantage of waste disposal by

landfill is that as the waste we generate mounts, the available land space shrinks. This is real food for thought.

3) **Incineration**: An incinerator is a large firebox in which refuse is piled, ignited, and allowed to burn at whatever equilibrium temperature naturally. The primary advantage of incinerators is that a significant reduction in the weight and volume of solid waste can be achieved, weight by up to 70 percent and volume by up to 90 percent—trading trash for ash. Ash disposal, however, certainly creates its own challenges.

The main disadvantage of incineration includes air pollution by combustion gases (greenhouse gases) and the presence of non-combustible waste such as glass and cans at the normal burning temperature. Most modern incinerators are now designed to do more of waste recovery than just burning waste, as mentioned under the fourth R – Recovery above. This approach points towards the concept of waste utilization rather than mere disposal.

Glossary Of Terms

The need for a glossary of terms is to ensure that readers understand the packaging terms used in the book. It is an attempt to define, clarify, and communicate the meanings of the packaging terms.

Aluminium Foil - A rolled section of aluminium less than 150 microns (micro-metres) thick.

Annealing –Annealing process is a method of reducing internal stress in the glass by controlled cooling during glass container manufacturing.

Bags - A pre-formed flexible container generally closed on all sides except one side, which forms an opening that may or may not be sealed after filling.

Bar Code – This is a machine-readable symbol. The symbol value is encoded in a series of high contrast rectangular bars and spaces.

Barrier - It is the ability to stop or retard the movement of atmospheric gases, water vapour and volatile substances through a packaging medium.

Basis Weight or Grammage – The weight per unit area of a material.

Blank – A piece of material from which a container or part of a container will be made by a further operation.

Carton Board – A paperboard used for manufacturing folding cartons.

Crease – Line or mark made by folding any pliable material.

Cullet – Glass recovered from production rejects or consumer-recycling program, crushed, and added to the normal glass furnace feed.

DPI - Dots Per Inch.

Double Seam - The seam at which the can ends are attached to the can body.

Film – Generally used to describe a thin plastic material.

Flashes – Excess plastic material that is squeezed out between mold parts during moulding.

Flute – A corrugated board medium used in the construction of corrugated board.

Gauge – A unit of measure describing the thickness of a material from which containers are made.

Gob – The amount of molten glass sufficient to make one glass container.

Hermetic – Impervious to air, gas, or fluids.

Hygroscopic – Tendency of a material to absorb moisture from the atmosphere.

Laminate – A material made by bonding many layers of different materials together to form a single sheet.

Manufacturer's Joint or Lap – The joint made by the box maker in the manufacture of a corrugated container.

Neck Finish - It is the portion of the neck that carries the thread, lugs or friction fit members to which the closure is applied.

Opacity – The ability of a material to prevent the transmission of light.

Parison – A partially formed glass shape that will be blown into a full shape of a glass container or the extruded hot plastic tube that will also be blown into a plastic bottle.

Porosity – The characteristic of material that allows the free passage of air or liquid.

Preform – The preliminary shape that will subsequently be transferred to a blow mould for inflation to a finished container.

Score – The crease or cut made on a paperboard to facilitate bending, folding, or tearing.

Shelf Life – The expected time within which the quality of a product remains acceptable.

Stripping – Removal of trimmings or off cuts from die-cutted folding carton blanks.

Substrate – Any material on which some action such as printing, coating, and so on is being performed, e.g. film, paper, etc.

Tolerance level – It is the permissible deviation from the specified quality parameters.

Web – A paper film or foil or any other flexible material as it is unwound from a roll and passed through a machine.

Yield – The number of container units that can be obtained from a given weight of a material or the area obtainable from a unit weight of the material.

References

1. Soroka, Walter; *Fundamentals of Packaging Technology*, 3rd edn. (Naperville, Illinois: Publisher, Year).
2. Soroka, Walter; *Fundamentals of Packaging Technology*, UK 2nd edition. Published by IOP, UK.
3. Correspondence Course, *The Institute of Packaging*, UK (1972).
4. Robertson, Gordon I; *Food Packaging, Principles and Practice*.
5. Leonard, Edmund A; *Packaging Specifications, Purchasing and Quality Control* (4th edition)
6. Leonard Edmund A.; *Packaging Economics* (1980).
7. Hirsch, Arthur; *Food Flexible Packaging – Questions and Answers*.
8. *The Stability and Shelf-life of Food*, edited by David Kilcast and Persis Subramaniam.
9. Stewart, Bill; *Packaging Design Strategy*.
10. *Packaging, The Facts*, IOP, UK.
11. Paine, F.A.; *The Packaging Media* (1977).
12. Sacharow, Stanley and Brody, Aaron L.; *Packaging: Introduction* (1987).
13. Sacharow, Stanley; *A Packaging Primer* (1978).
14. Sacharow, Stanley; *A Guide to Packaging Machinery* (1980).
15. Stewart, Bill; *Packaging as an Effective Marketing Tool*.
16. Selke, Susan E.M.; *Understanding Plastics Packaging Technology*.
17. Guss, Leonard M.; *Packaging is Marketing* (1981).
18. Sola Somade and Tunji Adegboye; *111 Questions and Answers in Packaging Technology* (2009).

About the Authors

Sola Somade is an industrial/ analytical chemist, having received a BSc degree in industrial chemistry from the City University, London, and an MSc in analytical chemistry from the Imperial College, University of London. She worked for Unilever Nigeria Plc for twenty years, where she held several key positions in product development, quality assurance, packaging development, and packaging  buying. She was a diploma member of the Institute of Packaging, UK, and she is now a Fellow of the Institute. Sola is recognised by PIABC as an Accredited Packaging Professional (APkgPrf) and an IOM3 Approved Assessor. She is the co-author of 111 Questions & Answers in Packaging Technology, and the co-founder of Superior Packaging Consultants Ltd – a firm of packaging consultants and trainers based in Lagos, Nigeria.

Tunji Adegboye holds a BSc degree in chemical engineering from the University of New Brunswick, Fredericton, Canada, and an MS degree in operations research and statistics from the Rensselaer Polytechnic Institute, troy, New York, USA. He worked for both Unilever Nigeria Plc and Cadbury Nigeria Plc for a total of twenty-seven years where he held such key positions as quality assurance  manager, packaging development manager, and technical manager, among others. He was a diploma member of the Institute of Packaging, UK, and is now a Fellow of the Institute. Tunji is recognised by PIABC as an Accredited Packaging Professional (APkgPrf) and an IOM3 Approved Assessor. He is the co-author of 111 Questions & Answers in Packaging Technology, and the co-founder of Superior Packaging Consultants Ltd – a firm of packaging consultants and trainers based in Lagos, Nigeria.

Printed in the United States
By Bookmasters